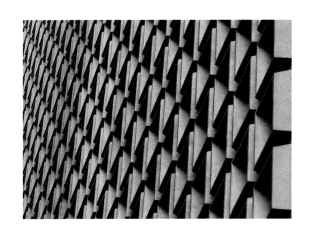

辉煌与毁灭
——现代建筑的另类解读

Splendor and Destruction
——An Alternative Reading of Modern Architecture

范建华 著
Fan Jianhua

云南大学出版社
YUNNAN UNIVERSITY PRESS

图书在版编目（CIP）数据

辉煌与毁灭：现代建筑的另类解读 / 范建华著. —昆明：云南大学出版社，2009
 ISBN 978-7-81112-924-3

Ⅰ. 辉… Ⅱ. 范… Ⅲ. 建筑艺术—研究 Ⅳ. TU-8

中国版本图书馆CIP数据核字（2009）第168366号

责任编辑：伍 奇 刘 雨
装帧设计：刘 雨

辉煌与毁灭
——现代建筑的另类解读
Splendor and Destruction
——An Alternative Reading of Modern Architecture

范建华 著
Fan Jianhua

出版发行：云南大学出版社
制　　版：昆明雅昌图文信息技术有限公司
印　　装：昆明富新春彩色印务有限公司
开　　本：889mm×1194mm　1/16
印　　张：11
字　　数：125千
版　　次：2009年12月第1版
印　　次：2009年12月第1次印刷
书　　号：ISBN 978-7-81112-924-3
定　　价：198.00元

社　　址：昆明市翠湖北路2号云南大学英华园内
邮　　编：650091
电　　话：（0871）5033244　5031071
E-mail：market@ynup.com

对于城市来说，建筑是辉煌与毁灭的墓志铭。

对于乡村来说，建筑是古典与孤傲的单行本。

For the town, architecture is the epitaph for splendor and destruction

For the country, architecture is the separate edition of classicism and detached arrogance

作者简介
Biography to the Author

范建华，云南人，学的是历史，研究的是现实，把视觉人类学作为自己生命的学科专业对待，于是走遍云南的山山水水和中国西部大部分民族地区：把南诏故地巍山和高黎贡山作为田野考察基地，一做就是十年；曾两次进藏，横上世界屋脊；穿越河西走廊，直过塔克拉马干大沙漠；环青海湖而感知黄河源头文明；放马内蒙草原而领略塞外风情；逐湘黔、访两广、探四川、问古徽州……也曾把田野考察触角放到国外，曾做客美国宾夕法尼亚州兰卡斯特的阿米什人家；还走访过圣达菲印第安部落；也曾在新西兰毛利人部落作过调查……总之，凡人类传统文明之地均想涉足。

由于曾任云南省社会科学院少数民族文学研究所所长，《山茶·人文地理》（现为《华夏地理》）杂志社社长，所以培养了用笔和相机思考学术、记录生活的习惯，出版过古城摄影作品《中国古城——巍山》、《中国古彝州——楚雄》、《视觉吴哥》和影视人类学调查实录《白族工匠村》、《滇藏文化带考察》（合著），在建筑摄影上有一点创作的心得，也有一点自己独特的理解和表达方式，因而这些作品都不能算是摄影家的艺术作品，只不过是一个人类工作者的视觉感悟而已。

Fan Jianhua was born in Yunnan Province, China, received formal training in history, and focuses his research on reality. Visual anthropology is his life work, having taken him to every corner of Yunna Province, as well as most of the ethnic regions in Western China. He spent 10 years doing fieldwork in Weishan, the birth place of the Nanzhao Kingdom, and the Gaoligongshan Mountain. He has traveled to Tibet twice, scaling the roof of the world; traversed the Hexi Corridor, cutting across the Takla Makan Desert; circled around Lake Qinghai, experiencing first hand the culture of the Source of the Yellow River; herded horses in Inner Mongolia, getting a taste of the land and culture beyond the Great Wall. In addition, he has visited the following provinces: Hunan, Guizhou, Guangdong, Guangxi, Sichuan, and Anhui. He has also done field research abroad: the Amish in Pennsylvania, the Pueblo Indians of New Mexico, and the Maori of New Zealand. His desire is to set foot in each and every locale that is significant to the development of the cultural heritages of humanity.

Fan was the director of the Institute of Minority Nationalities Literature at the Yunnan Academy of Social Sciences. He was the Editor-in-Chief of *Camellia* (currently *Huaxia Geography*), a journal of cultural geography. Experiences in the above fields have cultivated his skills in scholarly writing and photographic documentation. His published works in the field of photography are: *Old Chinese City-Weishan*, *The Old Yi Prefecture of China-Chuxiong*, *Visual Angkor Watt*. In the area of visual anthropology field documentation, he has published: *The Bai Village of Craftsmen*, *The Yunnan-Tibet Cultural Belt*. In architectural photography Fan has developed a personal insight, approach and expression. Therefore, the photos he has taken are not artistic works of a photographer, but simply the visual expression of the sense and sensibility of a working anthropologist.

关于现代城市的思考：范建华现代建筑摄影集序

郝光明

自从20世纪90年代开始，美国哥伦比亚大学美中艺术交流中心就与云南省诸多部门合作，从事文化以及生态保护方面的工作。范建华是对于我们的工作的最为坚定而明确的支持者之一。正因为如此，在范建华关于城市建筑的摄影作品即将公开展览并出版成书之际，中心愿以极大的热情和骄傲推举这一盛事。

中心在云南的项目虽然内容不一，但有一个主题始终贯穿于所有项目之中，那就是如何处理文化遗产与现代性，保护与发展之间的关系问题。这个问题是复杂又是无可回避的。范建华通过这本摄影集对这个问题的思考和讨论又做出了极有意义的增添。当然，建华此次的参与是通过影像的媒介而完成的，但是这部影集不应该被简单地看成摄影艺术，或者是民俗记录。它表述了一个经过仔细思考而呈现出来的世界观。

瓦尔特·本雅明（Walter Benjamin）在20世纪20年代曾经进行过一个宏大的研究项目，叫做"拱廊计划"（The Arcades Project）。这个计划的主旨是为了研究现代城市，建筑，资本主义以及人类的健康生存状态之间的关系。虽然计划没有完成，它仍然对于这一领域的知识建设贡献良多。本雅明的研究项目中有一个主要思想，反映在他对于资本主义的都市生活的梦幻转至噩梦般的描述之中。建华此次的摄影在一定程度上体现了与本雅明类似的思考。

本雅明在另一篇文章"关于历史哲学的思考"（Theses on the Philosophy of History）中批评了一个他认为未经严谨思考的历史观——所谓的历史的线性。这种历史线性认为历史事实是可以被确定并以清楚整齐的时间顺序呈现出所谓的"过去事项的真相"。基于对于历史唯物主义的独特理解，他的另类的历史观认为，我们应该"抓住在紧急状况下闪现出来的记忆"，用它来"爆破"（是在哲学意义上的"爆破"而不是实际意义上的）"当下"这样统一而空洞的时刻，为新的，未能确定的，在想象或者现实层面中的可能性开辟道路。这种可能性，被本雅明命名为"再生"。建华的摄影集中有很多照片正是见证了那种认为历史已经被规定好了，前进的"最佳途径"已经被找到了的傲慢的历史观。

本雅明对于僵硬秩序以及线性的结构的反感在雷姆·库哈思（Rem Koolhaas）的建筑论文中也有所反映。不过库哈思对于资本主义的过度发展比本雅明较为宽容。他认为"混乱"在城市的整体结构中是一个具有原创力的动力。他相信很多城市现在其实都按照一种他所谓的"灵活的都市主义"的方式运转。这并不意味着城市规划就不重要了，而是说城市规划是以一种灵活应变的方式运行，而不是完全基于总规划性的方式运行。在这一方面，库哈思与本雅明有暗合之处。两人都强调不可确定性，而不是可确定性。在我看来，库哈思最有意义的贡献在于他对于所谓"城市在经济及其他方面成为灵活的自我管理中心"的理念的阐释和推广。但是库哈思并不相信"绝对满足市场需要"的意识形态。他认为如果要真正理解一个事情，必须置身于其中。他在纽约"帕拉达"（Prada）旗帜店的设计就源于这种被他称为"参与式的批判"的理念。

珠江三角洲包括香港是令库哈思倍感兴趣的诸多地区之一。1996年他在此区开展了一个叫作"加度差异的城市"的项目。此地区除香港之外，还包括深圳、广州、珠海和澳门等。在这个项目中，库哈思一方面关注这些城市之间的差异，另一方面也注意到了它们在功能和定位方面逐步展示出来的共存的关系。这一观点体现了与受到"田园城市"影响的新城主义模式不一样的城市规划思维。"田园城市"的新城主义迄今依然无法完全克服其机械的拟人倾向以及一定程度上的具有乌托邦色彩的道德主义。库哈思把本地的环境放在第一位，由此削弱了在城市规划中呈现出来的普适主义倾向。他的观念虽然具有一定的争议性，却不失为对讨论议程必要的补充。

香港作为珠江三角洲的龙头是世界上少有的都市化程度极高的地方。要解读香港并不是一件容易的事。王家卫的电影在这方面做得非常出色。这位香港导演主要是通过记忆的碎片来理解香港的。这些记忆的碎片以一种既感性又抽离的方式描绘了这个城市的混杂性，它的瞬息万变的特色以及它的"混乱"。旺角、湾仔这些街区是计划与无计划的结合体，这一点最能够在街头的日常生活中得到体现，与库哈思所说的"灵活的自我管理中心"的理念相当吻合。

建华的摄影中体现了一种对于现代城市的爱恨交加的感情。在这方面，他与一系列著名的都市生活的观察者，比如马克思（Marx），韦伯（Weber），德克海姆（Durkheim），西姆（Simmel）甚至本雅明有共鸣。哈贝马思（Jurgen Habermas）把现代性看成是一个"未完成的计划"，需要依赖"主体的互动"来唤醒。这个观点也许有一定文化的局限性，但它的确是为现代性，包括现代都市生活的持续性提出了一个重要的讨论课题。简·雅各布斯（Jane Jacobs）对于这个问题可能具有最脚踏实地的回答。下面引用了一段她对于我所钟爱的纽约的下东区的描述：

"一条城市街道所传达的信任感是通过很多发生在路边的公共性的接触产生的。人们在一个酒吧停下来喝一杯啤酒，与卖菜的人交谈并得到某些信息，再把这些信息传达给报亭的小贩；顾客们在糕点店里交换意见，向两个坐在门口台阶上喝汽水的小孩子打招呼，这就是信任感。

这些行为表面看起来是很微不足道的，但是加在一起就一点也不微小了。这些在社区的层面上发生的看似随意的公共接触（任意的，与日常生活所需有关，在个人的掌控之中，而不是别人强加给他的）的总合构造了人们的公共身份，编织了一个公共尊重和信任的网络。这是一种资源，供个人或社区在需要的时候汲取和利用。"

在这一段文字中，雅各布斯提出了一个有关阶级的观念，比马克思的纯粹政治经济性的阶级观念（基于在马克思身后的阐释）更加广泛。这一段引文所传达的是一种由于共享空间和共同的利益而产生的多元的社区意识。

不管怎样，现代都市是一个未完成的计划；我们希望它的未来是不确定的。

Thoughts on the Modern City:
Preface to Fan Jianhua's Photography on Modern Architecture

Ken Kwan Ming Hao

Since the 1990's, the Center for United States-China Arts Exchange at Columbia University has been working with its many partners in Yunnan Province, China on cultural and biological conservation. Among all, Fan Jianhua has been one of the most ardent and articulate supporters of our work. Therefore, it is with great pleasure and pride that the Center heralds the publication and showing of Fan's photographic rumination on modern architecture.

Disparate though are the numerous projects carried out by the Center in Yunnan, the binding theme has been and remains to be the difficult but unavoidable issue of the relationship between heritage and modernity and between conservation and development. With this collection of photographs, Fan Jianhua has contributed significantly to the dialogue. To be sure, Fan's latest contribution is in pictorial form, but it is not to be taken simply as photographic art, or ethnographic documentation. Rather, it is a text representing a carefully thought out world view.

Walter Benjamin in the 1920's started an ambitious project called the Arcades Project to study the relationship between the modern city, architecture, capitalism and human well-being. Although unfinished, it remains one of the most important contributions to the field. A central thesis of Benjamin's is the dream-turned-nightmare characterization of capitalistic urban life. One aspect of Fan's photographs reflects to a certain extent this observation of Benjamin's.

Benjamin in another essay, *Theses on the Philosophy of History* criticizes the linear, causal notions of history. He dismisses the notion that historical facts can be identified and organized into an orderly temporal sequence of "things as they really were." His alternative historiography, based on his own understanding of materialism, stipulates that one must "seize on a memory as it flashes up in a moment of danger" to blow up (to "explode" in a philosophical sense and not in any material sense) the homogeneous and empty time of the present, bringing forth new and undetermined (emphasis mine) imaginative and material possibilities, which he sometimes called "afterlife". Reflected in many of Fan's photographs are material evidence of the arrogant belief that history has been defined and the "best" way going forward found.

Benjamin's antipathy towards order and linearity is reflected partially, in architectural terms, in the writing of Rem Koolhaas, although Koolhaas is more "forgiving" than Benjamin on issues of capitalistic excesses. Koolhaas sees chaos as a creative force in the overall makeup of a city. He believes that many cities nowadays operate on what he calls "improvised urbanism". It's not that urban planning has become irrelevant; it's just that planning often is done more spontaneously than in a deliberate, master plan-oriented way. In this, Koolhaas coincides with Benjamin—both emphasize undetermined-ness as opposed to determined-ness. What is most relevant, in my view, in Koolhaas' theory is his advocacy of and believe in what he calls "cities as centers of flexible self organization and improvisation, economic and otherwise." But Koolhaas is certainly not a "whatever market wants market gets" ideologue. He believes that in order to understand something, one must be part of it. His design of the Prada flagship store in New York City is based on what he calls "participatory criticism."

Among many of Koolhaas's interests is the Pear River Delta Region in China, including Hong Kong. In 1996, he initiated a project on this region, which in addition to Hong Kong consists of Shenzhen, Guangzhou, Zhuhai, and Macau etc. The project was called City of Exacerbated Differences. Koolhaas's focus was on the cities' extreme differences on the one hand and on the other, their evolving coexistence in functionality and identity. This represents a different understanding of urbanity from that of the Town-Country-inspired new urbanism, which has not been able to overcome entirely its anthropomorphic bent or rid itself of a certain degree of utopian moralism. Koolhaas's foregrounding of the local context and softening of the universalistic imperatives in urban planning, although controversial, is nevertheless a necessary addition to the agenda.

Hong Kong, the anchor of the Pearl River Delta Region, has few rivals in the world in urbanity. It's not an easy task to interpret Hong Kong. Wong Kar-wai's films do as good a job as any. The Hong Kong film director makes sense of Hong Kong mainly with memory fragments, which capture the hybridity, transience, and "chaos" of the city in a visceral yet cool-eyed way. Evident in neighborhoods such as Wan Chai and Mong Kok is a combination of the planned and the unplanned, mostly expressed in the every-day street life, which reflects quite pointedly Koolhaas's "self organization and improvisation."

A mixed emotion towards the modern city is an underlying theme of Fan's photographs. In this he joins a long list of distinguished observers of urban life—Marx, Weber, Durkheim, Simmel, and, yes, even Benjamin. Jurgen Habermas's characterization of modernity as "an unfinished project" to be revived on the basis of intersubjectivity is perhaps too culturally confined, but it does raise the important question of the continued viability of modernity, including modern urban life. Jane Jacobs has perhaps the most grounded response to this question. Here is a quote of her description of the Lower East Side of New York City, a neighborhood near and dear to my own heart.

"The trust of a city street is formed over time from many, many little public sidewalk contacts. It grows out of people stopping by at the bar for a beer, getting advice from the grocer and giving advice to the newsstand man, comparing opinions with other customers at the bakery and nodding hello to the two boys drinking pop on the stoop……

Most of it is ostensibly utterly trivial but the sum is not trivial at all. The sum of such casual, public contact at a local level—most of it fortuitous, most of it associated with errands, all of it metered by the person concerned and not thrust upon him by anyone—is a feeling for the public identity of people, a web of public respect and trust, and a resource in time of personal or neighborhood need."

In an interesting way Jacobs here provides a broader conception of the Marxian concept of class, which focuses strictly on the political economy (at least in what it has become). What the quotation illustrates is a multi-faceted community consciousness born of shared space and interest.

Warts and all, the modern city is a project unfinished, with a future, hopefully, undetermined.

图片检索　001~006
Image Search　001~006

前　言
辉煌与毁灭——现代建筑的另类解读　001~002
Foreword
Splendor and Destruction—An Alternative Reading of Modern Architecture　001~002

作者感言　001~157
Author's Reflections　001~157

后　记　158
Postscript　158

图片检索
Image Search

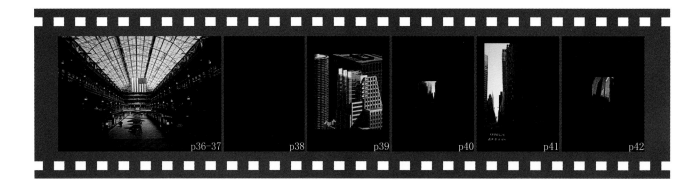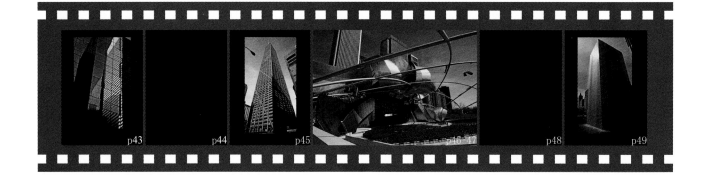

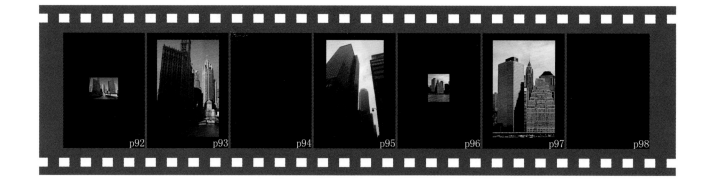

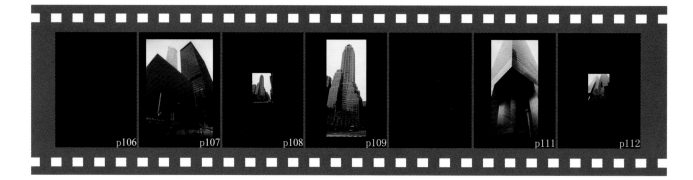

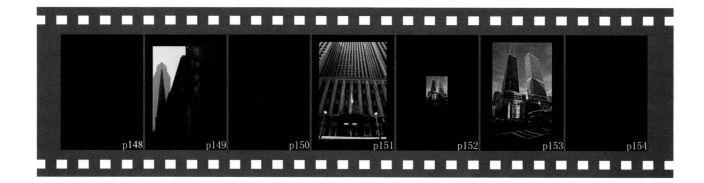
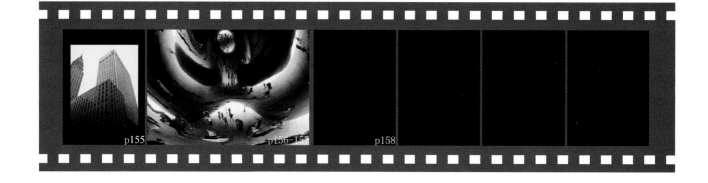

前　言

辉煌与毁灭——现代建筑的另类解读

　　玻璃、钢筋、混凝土是构成现代建筑的三大基本要素，亦被视为工业文明的象征性符号。出版这样一本现代建筑摄影集，本意并不是要展示现代工业文明的视觉张力、现代科技进步的智慧和现代建筑设计的创新能力，而是试图透过这些极具张力的建筑去解读隐藏在作品后面的哲学思辨。

　　笔者多年来对生态环境、经济社会、民族文化怎样实现协调和可持续发展进行的研究，受到纽约、芝加哥、多伦多、东京、香港等地现代建筑的巨大震撼而萌生了以摄影作品集的方式来诠释自己对农业文明——工业文明——后工业文明和生态文明的一种理解，进而阐释当下人与自然如何走向谅解与和谐的未来之路。解读现代建筑，使我们深知，现代建筑既表现了工业文明的巨大辉煌，又因之带来了无穷的毁灭。

　　仁者见仁、智者见智，在能平等对话的今天，你可以不同意我的观点，但你无权不让我表达我的观点。出版作品没有任何目的，如能让你感受到点什么，能引起一点不同理解，或引起一点共鸣，那就聊以自慰了。

Foreword

Splendor and Destruction—An Alternative Reading of Modern Architecture

Glass, reinforcing steel and concrete are the three fundamental elements of modern architecture. They are also looked upon as the defining symbols for industrial civilization. The primary intention in publishing this photographic book on modern architecture is not to show the visual tension of modern industrial society, nor the wisdom of modern technological advancement, nor the innovating capacity of modern architectural design, but to explicate the philosophical thinking hidden behind these structures of intense formal tension.

The many years that I have spent on researching the possible ways of establishing harmonious relationships between ecology and the environment, economy and society, and culture and ethnicity, as well as their sustainable development, have taken me to places such as New York, Chicago, Toronto, Tokyo, Hong Kong, etc. The modern architecture of all these locales has impacted me tremendously and brought forth the idea for a book of photography to expound my own understanding of agrarian civilization, industrial civilization, post industrial civilization, and ecological civilization and to explicate how man and nature at this moment in time can move toward reconciliation and onto a path leading to a future of harmony. We know well from reading modern architecture that on the one hand it reflects the grand splendor of industrial civilization, but on the other, it also brought with it endless destruction.

On the level communicative field of today, people hold differing views; you can disagree my perspective, but you have no right to stop me from expressing my views. I have no specific agenda in publishing my pictures beside the inner satisfaction that I'd feel if you are somewhat touched by the book, and a differing, or converging, perspective is engendered.

100多年前，霍德华在《明日的田园城市》序言中说："城市和乡村都各有其优点和相应缺点，而城市—乡村则避免了二者的缺点……这种该诅咒的社会和自然的畸形分隔再也不能继续下去了。城市和乡村必须成婚，这种愉快的结合将迸发出新的希望、新的生活、新的文明。"以往，城市与田园很难在一个层面上对话，今天我们遵循着历史文脉去寻找建筑的张力，会发现那些引起我们感动的视觉瞬间，已经超越了城市与乡村的界限。

Over a hundred years ago, Ebenezer Howard said, in the Introduction to his *Garden Cities of Tomorrow*, "…the chief advantages of the town and the country are set forth with their corresponding drawbacks, while the advantages of the town-country are seen to be free from the disadvantages of either." Town and country must be made one; and through this happy union, new hope, new life, and new society will be generated. In the past, town and country had no common platform on which to communicate. Today, in our heritage-based search for architectural splendor, we discover that in those fleeting moments of epiphany the boundary between town and country is transcended.

原片

原片

人行道不知所终，快车道让城市伤痕累累。当城市已经处在一个有很多立交桥、宽马路、复杂交通枢纽的状态下，我们很容易在喧嚣的城市气场中迷失自己。亚里士多德说，人们之所以从乡村来到城市，是为了生活得更美好。多少年后，人们用自己的行动打破了这种美好。向往田园牧歌般的宁静，不断激起人们迁居乡村的冲动，而当打着"回归自然"旗号的人们来到乡村时，大自然又将面临着新一轮的被改造和被解构。

The sidewalks go on forever; high-speed vehicular lanes scar the city lash after lash. With the innumerous elevated highways, the wide streets, the complicated traffic patterns, we get lost very easily in the resulting boisterous atmosphere. Aristotle said that people moved to the city from the countryside because they wanted a happier life. Years later, people smashed this happiness on their own volition. The longing for the serenity of rustic scenes and pastoral songs continue to arouse the urge to move to the country. However, when those bearing the banner of "Return to Nature" get to the countryside, nature will again be subjected to a new round of reformation and deconstruction.

原片

原片

原片

什么是传统，什么是现代，什么是未来，这是一个永远也无法回答的话题。昨天的当下，就是今天的过去；今天的现实，却是昨天的未来。一个无法逃脱的时光怪圈，决定着一个文明认知的思维怪圈。概念并不重要，重要的是珍惜自然赐予我们的每一滴净水、每一丝阳光、每一袭清风。

What is tradition, what is modernity, what is future, these are questions that can not be answered. Yesterday's present is today's past, but today's reality was yesterday's future. An inescapable and uncanny space-time matrix determines the unfathomable framework of cognitive thinking of an era. Concepts are unimportant; what is important is to treasure every drop of pristine water, every ray of sun light, and every wisp of light breeze that nature bestows upon us.

原片

原片

极目天地，反省古今，是大自然给予万物生机与活力，给人类一个美丽的家园。只有当我们自尝苦果之后，才会真正明白人与自然的关系，我们才会真诚地关爱自然，并同时关爱人类自身。今天我们终于清醒地认识到：自然不属于人类，而人类属于自然；对自然，永不再说"征服"二字。只有"和谐"才是人类得以生存与发展的前提。

From the farthest reaches of space and time to the here and now, nature has bestowed life and vitality, providing humanity with a beautiful home. Only after we've tasted the bitter fruit, can we then truly understand the relationship between man and nature and sincerely care for nature, and care for humanity itself. We have finally sobered up and recognized that nature does not belong to humanity, whereas humanity does belong to nature, and never will the word "conquer" ever be uttered against nature again.

"Harmony" is the only heading under which humanity can survive and progress.

原片

仰望一座座摩天大楼，我们常常可以指出它们所在的国家、城市和街区。然而在千篇一律的雷同与模仿中，这些"地标"逐渐趋于孤独，它们的躯体因为文脉带来的温度消失殆尽而逐渐冰冷，它们扮演的角色因为城市机器的飞速运转而逐成过客。

Looking up at the arrayed skyscrapers, we can often identify the country in which they are located, as well as the city and neighborhood. However, in their utter sameness and derivativeness, these "landmarks" gradually grow solitary; their bodies gradually turn ice-cold as the warmth provided by cultural heritage dissipates completely; and their role changed to that of a transient visitor in the face of the relentless churning of urban machinery.

在黄昏的影像里，摩天大楼的玻璃窗折射出街道上车水马龙的行色匆匆；公共广场的玻璃钢浮雕映射出芸芸众生的一脸茫然；拘谨的城中河倒映着高大建筑的千篇一律；华灯初现的橱窗里展示着流行文化前沿的物质符号；万家灯火的公路上疾驰的汽车拉成长长的光柱……黄昏城市的影调里，流露着人们不断地对自然进行沙哑的注解，对自己的心灵进行无力的颠覆与重构。

原片

Images of dusk reflected. The endless stream of harried travelers and vehicular traffic refracted through the glass facade of the skyscrapers; the perplexed expressions of the myriads of people reflected in the fiber-glass sculptures in the pubic plazas; and the unchanging sameness of the tall buildings reflected in the rigid city river. The shop windows with their evening lights just turned on, display the material signs of the avant-garde of pop culture. The ubiquitous headlights of the automobiles speeding on the highway formed into a light column extending farther and farther…. Revealed in the captured tone of the city at dusk are the incessant raspy-voiced annotation of nature and one's feeble attempt at subverting and reshaping the soul.

原片

原片

从农业文明走来，我们欣喜若狂地拥抱工业文明，物质财富的极大丰富性、力量和权势的表征，使得我们以为世界已达到辉煌的顶端，然而我们的愚昧恰恰在于丝毫没有意识到每一次科技的重大发明、每一次征服自然的巨大成功后面，却都暗含着一次次毁灭自我的玄机。

原片

Departing agrarianism we embraced with unbridled passion industrialism. Convinced that with its abundant material riches, symbols of strength and power, industrialism has propelled the world to the apex of its brilliance. Yet, we are ignorant for failing to recognize that every major scientific invention, every major nature-conquering success is accompanied by an insidious mechanism of self destruction.

摆在我们面前的实际情况是：在推动经济快速发展的进程中，往往我们是以摧毁文化和破坏生态为代价的。而我们所谓现代化的价值评判标准则完全是以工业文明作为参照体系和价值尺度。这样的结果就只能是在轰隆隆的机车声中传统的文明和脆弱的生态被无知又无情地碾碎。

原片

The reality we face is that in the process of promoting rapid economic development, we often use for payment the destruction of culture and wrecking of the environment. Moreover, we rely entirely on industrialism for reference and ethical parameters in establishing the criteria for assessing the value of modernization. The only outcome possible from this is the ruthless and ignorant shattering of traditional civilization and the fragile ecosystem amidst the loud rumbling of motors and machines.

原片

原片

人类从蒙昧时代、农业文明、工业文明、后工业文明一步步迈向生态文明的历史，是一个认识、感悟人与自然相互依存、相互遵守自然法则铁律的过程。虽然我们曾经犯过许多错误，然而上帝是允许人改过自新的，所以地球还没毁灭，人类尚未终结。

From pre-civilization, to agrarianism, to industrialism, to post-industrialism, humanity gradually strides towards an ecological civilization. This process of cognition and sensitization stipulates that man and nature must be symbiotic and mutually accountable to the iron-clad laws of nature. We have made many mistakes, but God allow humanity the opportunity to turn over a new leaf; therefore, the earth has not been destroyed and humanity has not been terminated.

原片

在冰冷的钢筋混凝土森林里行走的我，虽有灯红酒绿的喧嚣在耳畔回荡，虽有风驰电掣的车辆擦肩而过，然而看不见星空的夜晚，总是那样的孤寂与绝望。

原片

In the jungle of ice-cold reinforced concrete I walked, feeling invariably lonely and hopeless under the starless sky, as the gaudy neon din reverberates in my ears and cars whiz too closely by.

原片

原片

许多城市人为什么既不信仰宗教，也不信奉自然。因为他们从出生到终老，每天看到的是密密麻麻的街道，眼花缭乱的商品和不时传来的暴力、抢劫、杀人等罪行。在他们心中，上帝早已死去，而自然不知为何物。

原片

The reason that many city dwellers believe in neither religion nor nature is that from cradle to grave, they see before them thickets of streets, mind-boggling arrays of merchandise, and ever recurring violence, mugging and killing. Deep down, they believe that God is dead and nature is non-existent.

原片

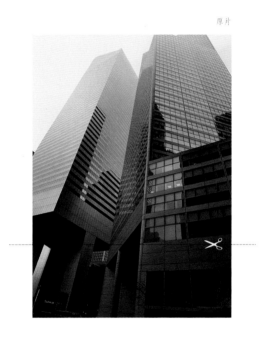

原片

当有一天我们在痛苦的工业文明巨大财富阴影下走出时，我们才确知一事：大地，不属于人；而人，属于大地。大地是我们的母亲；大地的命运，就是人类的命运，人若唾弃大地，就是唾弃人类自己。

原片

When we have come out from under the shadow of the great material wealth of the pain-ridden industrial civilization, we will know that the earth does not belong to humanity, whereas humanity does belong to the earth. Earth is our mother; the fate of earth is the fate of humanity. If humanity spurns earth, it spurns itself.

亨利·摩尔从来没有偏离过对生命的关注，在他的雕塑作品中，可以看到那些富有生命力象征的符号跃动在城市空间中。在城市罅隙仅存的绿地上，城市雕塑试图不改变大自然渲染的灵性去诠释和谐。这对大自然、对和谐刻意的注解，只是人们仓皇行程中一次短暂的心灵撞击，但依然让我们收获着浮华孤独中的宁静和感动。

原片

Henry Moore has never deviated from his care for life; in his sculpture pieces, one sees the life-energy motifs vibrating in the midst of urban space. In the rare urban greeneries, sculptured pieces attempt to explicate harmony without altering the intuition infused in nature, but this labored explication of nature and harmony is no more than a brief stirring of the soul on a harried journey, which nevertheless has given us serenity and fluidity amidst ostentatious solitude.

原片

原片

当今天工业文明的高楼大厦迅速吞噬着农业文明残存的花园乡野时，我们常常可以惊艳地发现许多高楼的屋顶悄然开辟出一个园囿，盛放着花与树，成为现代城市中的空中花园。然而相对于巴比伦时代，今天的空中花园却在孤芳自赏中早已发生了无奈的意义变迁。

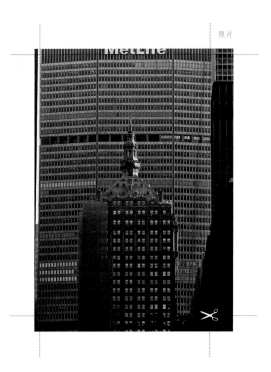

原片

原片

Today, as the skyscrapers of industrial society quickly devour the remnants of agrarian meadows and fields, we often discover to our pleasant surprise that many high towers have cultivated rooftop gardens, with vibrantly growing flowers and trees, becoming the hanging gardens of modern metropolis. However, in contrast with Babylonian times, these hanging gardens of today, in their solipsistic state, have long ago willy- nilly undergone a transformation in meaning.

当古老的教堂和现代化的高楼杂糅在一起，没有金属的质感，没有钢筋的迷茫，没有水泥的冰冷，更没有直指云霄的嚣张，在深邃的苍穹下，教堂更显示出她信仰的沉甸与宗教文化的沧桑。当唱诗班的旋律响起，人们极力虔诚地与上帝对话，从而掩饰心灵的孤独、思想的疲惫和对未来的恐惧，以获得假想中的涅槃。

原片

When aged churches and modern towers are juxtaposed, it is without the feel of metal, nor the perplexity of reinforcing steel, nor the iciness of concrete; and moreover, it is without the clouds-piercing arrogance. Under the expansive azure sky, the church reveals even more of her deep-rooted faith and weathered religious sojourn. As the melodic hymns ring forth from the choir, the parishioners with deep devotion converse with God, and thereby suppressing the loneliness of the soul, the fatigue of rational thought and the fear of the future, thus obtaining an imaginary nirvana.

原片

工业文明的出现，高度发达的生产力，使我们看到一个不争的事实：凡人迹所至，自然似乎都改变了模样。工业文明以来的经济增长所倡导的"人类征服自然"模式，使我们在发展中堕落，在科学中愚昧，其结果便是人与自然处于尖锐的矛盾对立之中，并不断受到大自然的报复与惩罚。掠夺——报复，再掠夺——再报复，成为200多年来人与自然关系往复不断的循环规律。

The emergence of industrial civilization, with its highly developed productive prowess, has shown us an indisputable fact: wherever humans set foot, nature seems to be altered whole-scale. The "Man conquering Nature" model advocated by the economic-growth oriented industrial society has made us decadent amid development, ignorant amid science, resulting in a poignantly contradictory relationship between man and nature, as nature incessantly punishes and retaliates against humanity. Plunder-revenge, plunder again-revenge again—this looping circle has been the unchanging pattern that characterizes the relationship between man and nature for the past 200 years.

后 记

本摄影集出版了，从视觉的角度，它一反我自己以往对古建筑摄影的特殊偏爱，投视和聚集点在现代建筑上，这也是对自己的一次否定之否定。

完成这一否定之否定的心理历程，看上去似乎有点格格不入，细想一下，还是本性使然。对老东西的偏爱，对时尚的反感，已很难改变，这就是一个冥顽分子的执著。

当然这种顽固虽不太入时，但也颇有一批志同道合者，所以它才得以出笼。这里要感谢一直以来对我帮助很大的挚友郝光明教授、郭游教授、齐骥教授、石明教授，还有李志旺教授、丁群亚教授、伍奇教授、刘雨教授，他们中有的暂没有教授的头衔，如齐、伍、李、刘诸先生，但他们的艺术修养和学识，远胜许多所谓教授，所以我先从心底封他们为教授，是他们给我思想的灵光、具体的帮助。

Postscript

This book of photography is published. From a visual perspective, it is a reversal of my past preference for old architecture. In this new volume, my gaze and focus is on modern architecture. It is in a sense a negation of a self negation.

The psychological journey undertaken for the negation of negation seems to be incongruous. On closer examination though, it is just inertia of one's nature. Love for the old, antagonistic to the trendy, these are attitudes that are hard to alter. Yes, this is the persistence of a fool.

To be sure, this type of obstinacy is not fashionable, but there nevertheless is a group of fellow travelers who made the publication possible. I want to thank Professor Ken Kwan Ming Hao, a close friend who has been persistently helpful, Professor Guo You, Professor Qi Ji, as well as Professor Li Zhiwang Professor Ding Qunya, Professor Wu Qi, Professor Liu Yu and Professor Shi Ming. Some among them are temporarily without the formal title of professor, for example. Qi, Wu, Li and Liu, but their artistic accomplishment and scholarship outshine many so-called professors. Therefore, from the bottom of my heart, I confer upon them the title of professor. They gave me the inspiration as well as the practical help.